Ham Radio:

The Beginners Ham Radio Guide On How To Set Up A Ham Radio

Table of content

Introduction

Since the evolution of mankind, mankind has been busy in inventing new technology for their own ease and comfort. In every domain of life humans are seeking to develop novel technology which is obviously better than its previous invention.

With expansion of cities and communities around the world humans had felt a need to develop communication between cities for trade or business. In early days man is send as messenger so he would have to travel hundreds of miles just to deliver a message which is time consuming and hectic thing to do so researchers from around the world started thinking about solving this issue and hence the milestone called radio was achieved.

The journey starts with an Italian scientist Guglielmo Marconi who gave the idea of signal transmission by practically demonstrating a telegraph sent from long distance over a radio signals in late 19th century. Later on in early 90's concept of radio was introduced and soon it became main source of news broadcasting and communication between long distances.

From idea of radio a thought emerged about setting up a radio anywhere you want. In late 90's radio was used for entertainment, news and other purposes so with variety of things it can deal in, it gives a pleasure to number of people to operate, setup and develop their own radios in their place. This concept gave birth to a concept called "Ham Radio "The first ever occasion in which ham radio was practically used is Christmas Eve of 1906 and a Christmas story was broadcast named as "silent night" in Bible.

Chapter 1 – Overview of Ham Radio and High Frequency Communications

Ham radio:

Amateur radio are also called as ham radio and beginners in radio operation are referred as hams. It can be adopted as a hobby as it was a trendy hobby in late 90's in USA. Even today we can see a lot of people using ham radio on commercial and individual level.

Hams and their interest:

Ham radio attract practitioners with a good vary of interests. Several amateurs begin with a fascination of radio communication which then develops into a hobby. A number of the focal areas hams adopt radio contesting, radio broadcast study, public service communication (PSC), technical experimentation and laptop networking.

Ham radio operators use numerous modes of transmission to speak. The two most typical modes for voice transmissions are:

- Area unit modulation (FM)

- Single sideband (SSB)

FM offers prime quality audio signals, whereas SSB is healthier at long distance communication once information measure is restricted. The amateur radio service (amateur service and amateur-satellite service) or service in setting up a ham is established by the International Telecommunication Union (ITU) through the International Telecommunication rules. National governments regulate technical associate in nursing the operational characteristics of transmissions and issue individual stations licenses with a unique decision sign. These beginner operators are tested for his or her understanding of key ideas in natural philosophy and also the host government's radio rules.

Radio amateurs use a spread of voice, text, image, and information communications modes and have access to frequency allocations throughout the RF spectrum to alter communication across a town, region, country, continent, the world, or maybe into area.

Every year, hams round the world participate in an occasion designed to bolster the power to setup and operate a good station victimization emergency power and temporary antennas, throughout a weekend long event referred to as Field Day.

High Frequency:

It is basically range of electromagnetic waves or radio waves between ranges of 3 to 30 MHz, it is used to provide noise free channels that enables smooth communication between transfer and receiver.

High frequency Communications:

High Frequency spectrum is used in molding communication system that benefits physical properties of the HF Radio Channel.

Using high frequency for communication:

- It's greatest worth is its ability to produce reliable short AND long-range on the far side Line Of Sight (BLOS) communications.

- It is reliable and needs little infrastructure to build up.

- It supports point-to point and point to multipoint conversations with less likelihood of noise in channels.

Ham radios and communication:

Amateur radio is independent of terrestrial facilities that may fail. It is distributed throughout community and there are no "choke points" or noise in channels like radio telephone sites.

Ham radios can be utilize for functions of non-commercial exchange of messages, wireless experimentation, self-training, non-public recreation, radio sport, contesting, and emergency communication.

For example you are on a trip or vacation with your school and to add some fun and excitement you can setup your own ham radio on spot and you can broadcast messages or song dedication etc. on your radio which will enlighten up the environment indeed.

More over ham radios can also be used as a key in disaster management. In case of any disaster or emergency you can set up a ham radio to boost up communication between victims and helpers.

An important example of "ham radio" is its use by New York City agencies to keep in touch with each other after their command center was destroyed during the 9/11 incident. Ham radio also came as a tool for disaster management during natural disasters like Hurricane Katrina, where all other communications failed, and also in the devastating flooding in Colorado in 2013.

There are few Intimidations that could be faced by hams or ham radios:

Attempted spectrum grabs, wherever enticing bands allotted to amateurs become vulnerable by industrial and alternative interests that need the precious spectrum, sporadically rear their head. Bands that are in demand, the 70cm band, was recently granted a keep of execution. Broadband over Power line (BPL), associate unrelated future technology that uses utility lines to supply web access, has the potential to cause interference and trim back from Amateur Radio Ham signals, in line with Major Ham cluster ARRL. There square measure people who dispute this.

Chapter 2 – Key Concepts of Ham Radio

Hams:

The operators of ham can also be called as hams. There is no specific qualification required for being ham but it requires some skill and knowledge of radio along with keen interest in this domain. Ham Radio operators are missionaries, celebrities, students, doctors, politicians, truck drivers or regular folks hence they can be people from all fields of life. There are number of legal talented young ham Radio operators currently than it was in previous times.

Ham radio significance:

Hams or Ham Radio is the kind aptitude to speak across the road, country or even worldwide, or perhaps with satellites, space stations in space or individual level. Even if there is no electricity, and also the cell phones and landlines are not working, with few equipment like a wire, a radio, and a battery, ham radio is there to use. Ham Radio enables North American nation relish their life-long distant friends, and an active technical activity or education. It boosts up the motivation, resources and encourages to experiment and play with new equipment's. It also encourages ham to adopt new styles that are governed by novel technology and newest way of communicating things.

You can't say that the sky is that the limit once there area unit footprints on the moon! You'll be able to go anyplace you wish, without boundary lines, and amateur radio will assist you get there!

Some major concepts that has to be kept in mind while setting up or talking about ham radio are:

- Skills that should be possessed by operators. They should have keen interest in ham radio along with particular set of skills.

- You should have license to operate radio in any place you want. Most hams have United States license and its validity last for 10 years.

- The next important thing is antenna, its setting location and many other factors concerning its performance.

- Purpose of ham is also an important aspects. You can either use it for personal or commercial purposes and you can also use it in emergency cases or in medical relief camps etc.

Hams square measure at the leading edge of the many technologies. They supply thousands of hours of volunteer community and emergency services once traditional communications go down or square measure overladen. All of them fancy being creators, not simply shoppers, of wireless technology.

Types of Ham Radios

There are three types of ham radios:

- **Handheld:**

Small and light-weight, hand-held transceivers enable ham operators to speak on the go; as a result of their low power output, hand-held ham radios generally have variety of solely five miles at the most; their range may be inflated by a close-by ham radio repeater.

- **Mobility:**

Designed to be used from associate operator's vehicle or a ham shack; their 200-mile vary is usually as a result of the raised power out there through either a home wall association or by an association through a vehicle's battery via a 12-volt outlet or power adapter.

- **Base station**

Designed to be used from associate operator's vehicle or a ham shack; their 200-mile vary is sometimes as a results of the raised power out there through either a home wall association or by an association through a vehicle's battery via a 12volt outlet or power adapter.

Some fun facts about ham radio:

- There are approximately a total of 50,000 radio hams in North American nation.

- There aren't any age or status constraints applying to people who consent for Canadian amateur radio license.

- There square measure Amateur Radio clubs attending nearly each group.

- Morse code is not any longer needed so as to get Basic Amateur Radio Certification.

Multi domain purpose:

- **Education:**

Self-education. It polishes your skills for broadcasting. It can also be adopted as field of education. Helping another pursuing knowledge and skills, are traditions in Amateur Radio. Once you have legal rights and authorization you never stop learning.

- **Intercommunication:**

Hams exercise communicating in groups or "networks" and individually on a regular basis to acquire practical expertise in interacting "traffic" under all circumstances.

- **Excellent Platform for all amateurs:**

It provides an efficient interactive system for hams around the world to contact each other and gather whole hams community on one common platform in friendly environment. This will boost up their knowledge and skills.

- **Public Service**:

It can be used for serving public in variety of ways. Ham radios offer communication during simulated or real emergencies, and for noncommercial occasions.

- **Experimentation:**

Within the framework of the Radio communication Act, amateurs with frequency allocations and license can experiment with new modes and techniques of radio communication. This will develop new methods and techniques for hams.

Chapter 3 – Tips to get Ham Radio License

Since radio instrumentation has the potential to interfere with alternative radio transmissions, operators should learn some info concerning radio, however it works, and therefore the rules governing amateur radio. There accustomed be six license categories within the North American nation, however that has a lot of recently been reduced to three: Technician, General, and Extra. Every future category needs a lot of data and grants additional privileges with regards to permissible transmission bands.

So the first thing you need to do before going to broad cast your channel or something on ham radio , you need to get your license and you should be aware of the terms and conditions for legally operating a ham radio. Here I have discussed few tips that would guide you how to get your license.United States license is worldwide known so I will be discussing about its classes and pre requirements in getting your license.

Ham radio license in United States

In US, amateur radio licensing is authorized by Federal Communications commission (FCC) underneath strict federal guidelines. Licenses to operate newbie stations for non-public usage is granted to individuals of any age when they show an interest in understanding of each pertinent of FCC policies and expertise of radio station operation and safety concerns.

Candidates as young as five years antique have exceeded examinations and have been granted licenses. December 2012 marked one hundred years of beginner radio operator and station licensing with the aid of the United States authorities.

Operator licenses are divided into distinctive training, each one of which corresponds to a growing degree of expertise and corresponding privileges. With new era of technology, the info of the instructions have modified drastically, leading to the modern gadget of open instructions. License validity last for 10 years and then operator have to renew his license.

License classes in US

The FCC classifications of licensing have evolved notably because of the software's inception. Whilst the FCC made the maximum changes recently. It has allowed positive existing operator training to stay beneath a grandfather clause. Those licenses would not be issued to new candidates, however present licenses may be changed or renewed indefinitely after period of 10 years. More over if any individual other than us citizen who wants to have US amateur radio license can appear in an exam conducted by volunteer examiners. It is conducted on monthly basis.

The two classes of ham radio are discussed below:

- **Novice classes**

The novice classes' magnificence operator license turned into for operators who had exceeded a five phrase in line with minute (wpm) Morse code examination and a basic idea exam. After the 1987 restructuring, privileges covered 4 bands inside the HF ranges 3–30 MHz, one band inside the Very High Frequency (VHF) varies from 30–three hundred MHz, and one band within the ultra-high frequency (UHF) ranges from 300–3,000MHz. This class became deprecated with the aid of the restructuring in 2000. novice operators received Morse code most effective privileges inside the entire Morse code and information simplest segments of the general magnificence quantities of eighty, forty, 15 and information and Morse code inside the general section of 10 meters in 2007 simply prior to the Morse code requirement.

- **Advance class**

The advanced magnificence operator license, whose privileges have intently resemblance with general class license, however covered 275 kHz of additional spectrum in the HF bands, changed into deprecated by the restructuring in 2000.

Ham radio types:

There are three types of ham radio. These are as follow:

- **Technician:**

The entry-level licensing alternative, giving access to any or all of the ham radio frequencies on top of thirty megahertz; these frequencies area unit found in North America specifically with restricted shortwave access to locations abroad.

- **General**

  The next grade from amateur radio licenses is that the General license and needs operators to pass the technician level license needs first; this license provides access to any or all amateur radio bands and sets the user up for worldwide communication.

- **Amateur further**

  The highest of all 3 licenses; needs users to pass each the Technician License take a look at and therefore the General License take a look at, giving the operator access to any or all operative privileges each abroad and within the U.S.

Chapter 4 – Tips to set an Antenna and set a Station

A radio antenna is the critical hyperlink in any receiving or transmitting station whether or not it is used for ham radio, quick wave listening, or for industrial or professional use. The general performance of the radio station depends upon the overall performance of the radio antenna. An efficient radio antenna will enable the overall performance of the whole radio communications station to be maximum, whereas a poor antenna will degrade the competencies of the transmitter and receiver no matter how correct they are.

Radio stations used for professional or business packages have a big share in reliable performance. They may site antennas in which they may give a suitable level of performance but for ham radio enthusiasts or people who have adopted it as a hobby and brief wave listeners it's far essential to put it in the high-quality antenna around the residence.

Only a few ham radio operators are able to utilize an area or other huge location, and frequently the radio antenna might be something of a compromise. However, via following some hints, it's far feasible to make the excellent of any radio antenna set up and make sure that its performance is as true as possible further to this it's far worth citing that it also includes important to adopt some experimentation to find out what type of radio antenna works satisfactory for a given place and fashion of radio operation.

A few pointers and guidelines are given below for your ease, but these can simplest be general pointers, and they're now not exhaustive. However they shape a terrific beginning place when deliberating installing a ham radio antenna machine.

General settings of an antenna for ham radio:

One of the most vital components of putting in any radio antenna is its vicinity. The region of the antenna will govern many aspects of its operation, and therefore the vicinity of the antenna ought to be determined alongside the kind of antenna to be used.

Some of factors associated with the antenna must be considered:

- Pick an area in which the radio antenna can "see" all around: in order for to operate at its best it have to be capable of "see" all round it. So that you can achieve this it need to be kept far from nearby objects that could act as a screen. In this manner the maximum amount of sign can be attain or go away the antenna without being absorbed in close by gadgets.

- Secondly keep in mind that nearby gadgets can "detune" an antenna: whilst thinking about the location of a radio antenna it is really worth remembering that close by gadgets can detune an antenna despite the fact that they do no longer have an effect on the all spherical visibility.

- Surrounded gadgets can purpose an antenna to operate away from its resonant factor and become much less green. It can be very critical for antennas which might be cut to a particular length and do no longer have a means of being tuned in situ. Many items can motive this to show up - steel gadgets in addition to electric wiring are mainly horrific however even trees can degrade the performance of antennas in this way. Typically the consequences are important inside distances of a wavelength or the nearer the object and the more the conductivity the greater the effect.

- Another fact keep in mind appropriate points for anchoring antennas: Horizontal antennas want anchor factors at either stop. This is worth thinking about whether there are any appropriate anchor factors already in lifestyles. Chimneys or different points on the residence can provide one suitable point.

- Trees will also be positioned with no trouble, even though pulley schemes are required to permit any motion in the tree because of wind to be taken up without snapping the antenna twine. Additionally it may be feasible to erect a pole or antenna mast and attention can be given to this possibility and its vicinity. Anything alternative is decided upon, this must be considered at the outset.

- Inner or out: frequently the use of an internal radio antenna may also need to be taken into consideration. Outside antennas function higher because they may be in addition faraway from gadgets on the way to introduce loss or detune the antenna. It is very tough to estimate the quantity of loss which having an antenna in the house has. The roof or brickwork will motive the sign to be reduced, in particular while it's far moist. The amount of loss can even depend on the frequency. For VHF and UHF indicators this can be a whole lot greater.

- Some other factors that should be observed while setting an antenna are antenna height, matching of antenna with feeder, safety aspects of antenna, consideration of feeders that enables maximum power transmitted by an antenna.

Chapter 5 – Learn how to set up a Ham Radio?

Owning and in operation a ham radio, or amateur radio, offers people a fun and rewarding hobby or perhaps how to speak throughout a crisis. Operators should make sure that their rig is setup properly, and that they should additionally become licensed for in operation a ham radio before use. Being knowledgeable within the use of a ham radio parades a world of prospects to operators, permitting them to speak with other people and even permits them to speak regionally throughout emergency situations.

To get the foremost use out of owning and in operation a ham radio, house, owners have to be compelled to perceive all of the choices and options that the communication system has got to provide. They ought to even be accustomed to the various sorts of ham radios, common brands, what the necessities square measure to work a ham radio, and places to seek out and get ham radios, like at native physical science stores or on-line marketplaces like eBay. So, to get pleasure from the system within the trendy age, it is important to be told concerning frequency sorts, sorts of licenses, parts of a ham radio, and therefore the advantages of in operation a ham radio.

There is an important aspect to preserve in thoughts. Ham radio is an exquisite food for many people who have adopted it as a hobby. There are nearby emergency infrastructures (which we're organizing up radios for in this sequence) besides this there are other complete-direction meals concerning extreme Frequency communications where low energy activists (like me) want to check that how many states round the sector will "work" in Morse code. Morse code isn't obligatory for a ham authorization any longer, but that's another thing.

Digital modes are also there. Some operators enjoy doing sluggish-scan TV – directing photographs to Europe or anyplace, as any other complete meal with the aid of itself. The "guides" in that one take account of virtual images, transmitter techniques, photograph-shopping, in addition to "ordinary" High Frequency radio usage.

This entry stage path is one step, but at hand are loads. There's a ham station involved the transnational area Station, and "running" Japan the usage of nothing extra than a Very High Frequency (VHF) radio and a hand-held directional small beam antenna, is a full "subsequent meal" for some.

Elmer:

In amateur radio, the term "Elmer" refers to a mentor, or to the act of mentoring others. This is often integral to the ham radio spirit of serving to others. Several more-experienced hams can volunteer their time to answer queries, give tutoring or teach categories to anyone desire to enter amateur radio ranks, or World Health Organization merely needs to be told additional regarding it.

The amateur radio community could be a various network of clubs and people like an expert in an exceedingly type of areas. For some, sharing or business enterprise sensible data is their approach of giving back to the ham radio community.

There square measure several engineering and scientific professionals concerned in ham radio still as several celebrities. Journalist director Cronkite, baseball pitcher Bokkos Swobada, TV/radio temperament Jean Shepherd, legislator Barry Goldwater, and musician Joe Walsh square measure among the thousands of notable personalities World Health Organization square measure or were hams. Honor winner for Physics Joseph Taylor is a fanatical ham, and has developed many modulation formats that square measure appropriate for very weak signal communications.

Components of Ham Radio:

- **Shake:**

It depicts the location from where ham radio would be operated

- **Antenna:**

It is used to catch signals

- **Cables:**

These are used to connect transceiver with antenna and to make other wired connections

- **Guide wires:**

These wires helps in stabilizing of antenna.

- **Power supply:**

Provides power to equipment's of ham radio

- **Fuses:**

It protects electric equipment from power damage.

- **Battery:**

It is used to give current supply to antenna and other equipment.

- **Transceiver:**

It is used to receive signals

- **Antenna tuner:**

It helps in improving antenna capacity to catch signals efficiently.

- **Antenna rotator:**

Rotates antenna in different direction so that signals are catch accurately.

Setting up a ham:

First, if you're a Maker, then you have already got lots in you not unusual with the ham radio network. Hams are tinkerers, developers, fixers, and inventors by means of nature. As an Entrepreneur in this field, the field (or building your own field) is not most effective allowed, it's advocated! Of the numerous radio services obtainable from business broadcasting to Commercial purposes to public protection.

Amateur radio is the handiest one in which gadget may be home made and tuned to any frequency or channel that hams have access to. Flexibility, experimentation, and hacking are a way of existence with hams. Ham radio has many aspects it's absolutely 1000 interests in a single. You can dive deeply into electronics, antennas, digital communications, public service, competitive working, solar and geophysics technology, global-extensive "DX-ing," or just use ham radio as a private communications tool.

A few hams attention on just one or a few topics while others try and enjoy it all! As a Maker, you're in all likelihood maximum interested in the electronics, however once you begin digging in, you never understand in which it would lead or where you could apply your talents.

Whilst they're at the air, hams use dozens of different sorts of alerts; some are everyday voice indicators wherein we certainly talk to each other and yes some hams use Morse code at the same time as others are designing their very own virtual protocols to send statistics and messages around the world. Hams have their very own e mail and data networks. There are even ham radio satellites that relay signals, such as a ham station at the worldwide space Station that the astronauts use — they're hams, too. Hams perform from their homes, their cars, and even from mountaintops and islands!

All that equipment sounds highly-priced, but it doesn't must be. much like beginning your own workshop, you may preserve it simple, purchase used gear, scrounge for elements and portions, and work with different extra experienced hams (we call the mentors Elmer) to get started. The only access-degree radios fee much less than a hundred dollar and you can build your personal antenna.

Loose software program is extensively available on the way to use some of the distinctive sorts of signals, even Morse code which hams talk over with as "CW" for continuous wave. So, a primary radio, a couple of cables, and also you're in commercial enterprise to start as a novice ham. Just get concerned, search for funny-looking antennas, and ask around. You'll be amazed how pleasant, fun loving and helpful a few hams are!

Conclusion:

Concluding the description I can say that Ham radios afford their operators the chance to get involved in with folks worldwide that they might unremarkably check with and share common interests and new concepts with others.

In associate degree emergency state of affairs, ham radios additionally permit their operators to speak with authorities to function a possible resource in those styles of things at intervals in their community. Ham radios area unit still helpful within the fashionable age thanks to their ability to still work even once different varieties of communication don't seem to be. Operators additionally fancy opportunities to move and even hold contests for contacting the foremost users in an exceedingly given period of time.

Adopting it as a hobby is another interesting way to get in touch with this fully loaded piece of invention. Even if a person is interested in electronics or media stuff he can begin his practical training by setting up a ham radio. It is indeed skill building practice.

For those that decide that in operation a ham radio continues to be helpful, they'll get the instrumentation at numerous locations, however ought to be bear in mind to get a license before causation any messages. Before shopping for a ham radio, bear in mind to analyze accessible choices and study the various styles of ham radios, the necessities to awfully operate one, and the way to shop for ham radios safely and firmly on eBay.

It's conjointly serving to build basic skills that aren't any longer tutored at school rising not solely our ability to speak in disasters, however adding back a number of the 'lost tools' that Americans accustomed be notable for the power to try and do things ourselves. You will find this text utterly useful and comprehendible. I am 100% sure that with your interest, knowledge and guideline from this text you can surely setup your own ham radio.

FREE Bonus Reminder

If you have not grabbed it yet, please go ahead and download your special bonus report *"Leptin Resistance. 21 Leptin Recipes For Weight Loss & Healthy Living"*.

Simply Click the Button Below

OR **Go to This Page**

http://easyweightlossway.com/free/

BONUS #2: More Free & Discounted Books

Do you want to receive more Free & Discounted Books?

We have a mailing list where we send out our new Books when they go free or with a discount on Kindle. Click on the link below to sign up for Free & Discount Book Promotions.

=> Sign Up for Free & Discount Book Promotions <=

OR Go to this URL

Made in the USA
Monee, IL
07 July 2026